Bibliographic information published by the German National Library:

The German National Library lists this publication in the National Bibliography;
detailed bibliographic data are available on the Internet at http://dnb.dnb.de .

Imprint:

Copyright © 2018 GRIN Verlag
Print and binding: Books on Demand GmbH, Norderstedt Germany
ISBN: 9783668993884

This book at GRIN:

https://www.grin.com/document/494142

Akinmayowa Adedoyin Shobo

The effect of drought and salinity on secondary meta-bolite of plants

GRIN Verlag

LITERATURE REVIEW

BY

2019. SHOBO, Akinmayowa Adedoyin

SALINITY AND DROUGHT: ITS EFFECT ON SECONDARY METABOLITES IN PLANTS

ACKNOWLEDGEMENT

My heart goes out to my supervisor, --------------------for his unflinching support to the development of this work. I cannot fully express my gratitude to all my lecturers of the Department of Biochemistry for their patience and professional approach towards the realization of this work. Finally, I say a big thank you to my parents and friends for their love and support.

SALINITY AND DROUGHT: Its effect on secondary metabolite in plants

TABLE OF CONTENT

4.7 Effect of drought and salinity on Secondary Metabolites

Conclusion

Reference

CHAPTER ONE

1.1   Introduction

It is widely accepted that food security around the world is decreasing due to detrimental effects of various unfavorable environmental conditions. To this effect, minimizing these effects continues to be top priority globally to ensure food security under changing climate.

Plants are frequently exposed to a plethora of conditions that may or may not be favourable to their optimum functioning. Some of conditions including micro-organisms, air, water, gases (O_2 and CO_2) among others contribute positively to the growth of plants (Gleadow and Woodrow 2002, Falk *et al.* 2007, Ballhorn *et al.* 2011). On the other hand, an excess or deficit of these environmental conditions may disrupt the normal and adequate functioning of plant which leads to the phenomenon of stress to the plant.

Generally, stress (with respect to plant metabolism) has been described as an adverse condition which inhibits the normal functioning and total well-being of a plant (Jones and Jones, 1989; Mahajan and Tuteja, 2005; Aksoy, 2008). In their environment, plants are vulnerable to numerous forms of stress yet it is known that under normal biological conditions, what constitutes stress for one plant may be optimum for another plant (Mahajan and Tuteja, 2005). All these stress factors (drought, salinity, extreme temperature etc.) collectively have been studied and known to be associated with negative consequences to the overall plant growth and productivity.

Cold, heat, salinity, drought (water-deficit condition), flooding (excess water), radiations (high intensity of ultra-violet and visible light), chemical and pollutants (heavy-metals, pesticides etc.), oxidative stress (reactive oxygen species, ozone), wind (sand and dust particles in wind) are categorized as *abiotic stress* while pathogen (viruses, bacteria, fungi), insects, herbivores, rodents are called *biotic stress*, as they are caused by plant's interaction with other organisms. Additionally, *abiotic stress* in fact has been attributed to be the principal cause of crop failure worldwide, dipping average yields for most major crops by more than 50% (Bray *et al.*, 2000; Mahajan and Tuteja, 2005).

In the presence of a potential stressor (biotic or abiotic), certain cellular responses are initiated primarily by interaction of the extracellular material with a plasma membrane protein. This extracellular molecule is called a ligand (or a stressor) and the plasma membrane protein, which binds and interacts with this molecule, is called a receptor. The stressor is perceived by the receptors present on the membrane of the plant cells, the signal is then transduced downstream and this results in the generation of second messengers including (but not limited to) calcium, reactive oxygen species (ROS) and inositol phosphates Mahajan and Tuteja, 2005). This further leads to a cascade of signal transduction, followed by a physiological response such as an increase in the production of secondary plant products (Selmar and Kleinwachter, 2013).

1.2   Definition of Salinity

Conceptually, salinity is the quantity of dissolved of salt content of the water. Salts could also be described as compounds like sodium chloride, magnesium sulfate, potassium nitrate and sodium carbonate, which dissolve into ions. For example, the concentration of dissolved chloride ions is sometimes referred to as Chlorinity (Anonymous, 2018).

Some researchers have grouped saline environments broadly as being either wet or dry. Wet saline habitats tend to occur in coastal regions and are dominated by salt marshes. Since these areas border the sea, they are subject to periodic inundations, and as a result the level of salinity fluctuates over time. Dry saline habitats are usually located inland, often bordering deserts (Flowers 2004; Bado *et al.*, 2016). Other types of saline environments include seashore dunes, where salt spray is a salinizing factor, and dry salt lakes. Common features of saline environments are the salinity of the soil and/or of their associated water resources and specialised flora and fauna. The most abundant salts in saline soils are sodium chloride (NaCl) and sodium sulphate (Na_2SO_4), which may be associated with magnesium (Mg) salts (Bado et al., 2016).

Salinity is one of the most important abiotic stresses in arid and semiarid regions (Zhang et al., 2010; Sahu et al., 2011; Olfati et al., 2012; Talebi et al., 2015). Salt stress retards plant growth and yield, and has become a serious problem in the world (Horvath *et al.,* 2007; Kirdmanee, 2009; Moghbeli *et al.,* 2012).

The impact of salt stress has been correlated with some morphological and physiological traits such as reduction in fresh and dry weight (Chartzoulakis and Klapaki, 2000). In fact, salinity affects plant metabolism by disturbing their physiological and biochemical processes of plants due to ionic and osmotic imbalances which slows down plant growth and productivity (Munns, 2005). The deleterious effects of salinity on plant growth are associated with low osmotic potential of soil solution, nutritional imbalance, specific ion effect, or a combination of these factors (Ashraf and Harris, 2004).

1.3   How to measure salinity

Plant growth responses to salinity vary with plant life cycle; critical stages sensitive to salinity are germination, seedling establishment and flowering (Ashraf and Waheed, 1990; Flowers, 2004). Criteria for evaluating salinity tolerance in crop plants vary depending on the level and duration of salt stress and the plant developmental stage (Shannon 1985; Neumann, 1997). In general, tolerance to salt stress is assessed in terms of biomass production or yield compared to non-stress conditions. In conditions of low to moderate salinity, the production capacity of the genotype is often the most important measure, whereas survival ability is often used at relatively high salinity levels (Epstein *et al.,* 1980). The physiological mechanisms that play a major role in maintaining the production capacity of a genotype are not the same as those that contribute to tolerance at extremely high salt concentrations. Genotypes are generally evaluated using phenotypic observations. Phenotypic selection parameters include:

- **Germination:** Germination tests are easy to perform and may be important where the crops are required to germinate and establish in saline conditions. However, germination in saline conditions is not often associated with salinity tolerance in subsequent growth stages (Dewy, 1962; Shannon, 1985; Flowers, 2004).
- **Plant survival:** selection on the basis of plant survival at high salt concentrations has been proposed as a selection criterion for tomato, barley and wheat (Rush and Epstein 1976; Espstein and Norlyn 1977). The ability of a genotype to survive and complete its

life cycle at very high salinities, irrespective of yield potential under moderate salinity levels, is considered as being tolerant in the absolute sense.

- **Leaf damage:** Since most crops are glycophytes, they are unable to restrict toxic salt ions being translocated from roots into shoots and leaves. Consequently, salinity damage may be readily observed by leaf symptoms of bleaching and necrosis. Screening for salt tolerance by leaf damage is therefore common (Richards *et al.* 1987; Gregorio *et al.*, 1997).

- **Biomass and yield:** For plant breeders, yield and biomass are obvious parameters in assessing salt tolerance (Richards *et al.*, 1987). These parameters, however, do not provide information on the underlying physiological mechanisms. In the past, plant breeders have not been interested in physiological mechanism; that a genotype was tolerant was sufficient, the physiological mechanisms were regarded as academic. However, with the emergence of gene function studies, this view is changing.

- **Physiological mechanisms:** Physiological mechanisms that confer tolerance to salt may be harnessed for screening. These may include measurements of tissue sodium content, ion discrimination and osmotic adjustment. Surrogates such as carbon isotope discrimination which give a general indication of plant stress may also be used (Flowers and Yeo 1981; Pakniyat *et al.*, 1997).

1.4 Definition of Drought

Drought is one of the most significant environmental stressor that adversely disrupt plant physiology. It can be described as a temporal reduction of environmental moisture status relative to the mean state. Because of the complexity of drought, it is often studied only by separate aspects of the phenomenon (e.g. meteorological drought, soil drought, etc.) (Anonymous, 2018). In defining drought, it is particularly important to distinguish between dryness and drought. Dryness is a constant feature of an arid area caused by the climate. The total area of arid climates is estimated at about 42% of the Earth's land. Drought, on the other hand, is a temporary phenomenon related to the failure of usual precipitation. It always results in temporary loss of water and plant resources) (Anonymous, 2018).

Drought is often defined as a temporary situation when the water demand of a hydrological system (which may be an ecosystem or an anthropogenic system) exceeds the income of water from any sources.

1.4.1 Causes of drought

Drought is a complex physical and social process of widespread significance (American Meteorological Society, 2003; Owens *et al.*, 2003). Plant experiences drought stress either when the water supply to roots becomes difficult or when the transpiration rate becomes very high (Anjum *et al.*, 2011).

According to Hosseini *et al.* (2009); Drought severity is caused not only on the duration, intensity and spatial extent of a specific drought episode, but also on the demands made by human activities and vegetation on a specific region's water supply.

Based on their respective; drought may be categorized as meteorological, agricultural, hydrological, and socio-economic droughts (Hosseini *et al.*, 2009) which is discussed below.

1.4.2 Types of drought

Drought is considered the single most devastating environmental stress, which decreases crop productivity more than any other environmental stress (Lambers *et al.*, 2008).
There are two types of drought based on their several etiologies including:

Meterological Drought: This is caused by a continuous shortfall in precipitation is coupled with higher evapotranspiration demand leading to agricultural drought (Hosseini et al., 2009; Mishra and Cherkauer 2010). Precipitation is a major source of drought in most ecosystems.

Agricultural Drought: This is the lack of ample moisture required for normal plant growth and development to complete the life cycle (Manivannan et al. 2008; Hosseini et al., 2009). It occurs when soil moisture is inadequate to meet the needs of a particular crop at a particular time.

Hydrological drought: refers to deficiencies in surface and subsurface water supplies (Hosseini et al., 2009).

Socioeconomic drought: occurs when physical water shortages start to affect the health, well-being, and quality of life of the people, or when drought starts to affect the supply and demand of an economic product (Moghaddas- Farimani and Hosseini 2004; Zamani et al. 2006; Hosseini et al., 2009).

1.4.3 Consequences of drought

The permanent or temporary water deficit has been demonstrated to severely hampers the plant growth and development more than any other environmental factor (Anjum *et al.*, 2011). Drought negatively impacts include growth, yield, membrane integrity, pigment content, osmotic adjustment, water relations, and photosynthetic process (Benjamin and Nielsen, 2006; Praba *et al.*, 2009). Drought stress is affected by climatic, edaphic and agronomic factors. The susceptibility of plants to drought stress varies in dependence of stress degree, different accompanying stress factors, plant species, and their developmental stages (Demirevska *et al.*, 2009). Acclimation of plants to water deficit is the result of different events, which lead to adaptive changes in plant growth and physio-biochemical processes, such as changes in plant structure, growth rate, tissue osmotic potential and antioxidant defenses (Duan *et al.*, 2007).

With respect to plant growth; drought can impair germination (Harris *et al.*, 2002). Cell growth is considered one of the most drought-sensitive physiological processes due to the reduction in turgor pressure. Growth is the result of daughter-cell production by meristematic cell divisions and subsequent massive expansion of the young cells. Under severe water deficiency, cell elongation of higher plants can be inhibited by interruption of water flow from the xylem to the surrounding elongating cells (Nonami, 1998). Drought caused impaired mitosis; cell elongation and expansion resulted in reduced growth and yield traits (Hussain *et al.*, 2008). Water deficits reduce the number of leaves per plant and individual leaf size, leaf longevity by decreasing the soil's water potential. Leaf area expansion depends on leaf turgor, temperature, and assimilating supply for growth. Drought-induced reduction in leaf area is ascribed to suppression of leaf expansion through reduction in photosynthesis (Rucker *et al.*, 1995). A common adverse effect of water stress on crop plants is the reduction in fresh and dry biomass production (Zhao *et al.*, 2006).

Khan *et al.* (2001) conducted a study comprising of six treatments, namely, control (six irrigations), five, four, three, two and one irrigation in maize. It was concluded that plant height, stem diameter, leaf area decreased noticeably with increasing water stress. The reduction in plant height could be attributed to decline in the cell enlargement and more leaf senescence in the plant under water stress (Manivannan *et al.*, 2007a). Drought led to substantial impairment of growth-related traits of maize in terms of plant height, leaf area, number of leaves/plant, cob length, shoot fresh and dry weight/plant. Furthermore, Kamara *et al.* (2003) revealed that water deficit imposed at various developmental stages of maize reduced total biomass accumulation at silking by 37%, at grain-filling period by 34% and at maturity by 21%.

With respect to plant yield; plant yield often is the result of the expression and association of several plant growth components. The deficiency of water leads to severe decline in yield traits of crop plants probably by disrupting leaf gas exchange properties which not only limited the size of the source and sink tissues but the phloem loading, assimilate translocation and dry matter portioning are also impaired (Farooq *et al.*, 2009).

Drought stress inhibits the dry matter production largely through its inhibitory effects on leaf expansion, leaf development and consequently reduced light interception (Nam *et al.*, 1998). Drought at flowering commonly results in barrenness. A major cause of this, though not the only one, was a reduction in assimilate flux to the developing ear below some threshold level necessary to sustain optimal grain growth (Yadav *et al.*, 2004). When maize plants were exposed to drought stress at teaseling stage, it led to substantial reduction in yield and yield components such a kernel rows/cob, kernel number/row, 100 kernels weight, kernels/cob, grain yield/plant, biological yield/plant and harvest index (Anjum *et al.*, 2011a). Drought-related reduction in yield and yield components of plants could be ascribed to stomatal closure in response to low soil water content, which decreased the intake of CO_2 and, as a result, photosynthesis decreased (Chaves, 1991; Cornic, 2000; Flexas *et al.*, 2004). In summary, prevailing drought reduces plant growth and development, leading to hampered flower production and grain filling and thus smaller and fewer grains. A reduction in grain filling occurs due to a reduction in the assimilate partitioning and activities of sucrose and starch synthesis enzymes.

With respect to photosynthetic activities; the ability of crop plants to acclimate to different environments is directly or indirectly associated with their ability to acclimate at the level of photosynthesis, which in turn affects biochemical and physiological processes and, consequently, the growth and yield of the whole plant (Chandra, 2003). Drought stress severely hampered the gas exchange parameters of crop plants and this could be due to decrease in leaf expansion, impaired photosynthetic machinery, premature leaf senescence, oxidation of chloroplast lipids and changes in structure of pigments and proteins (Menconi *et al.*, 1995). Anjum *et al.* (2011a) indicated that drought stress in maize led to considerable decline in net photosynthesis (33.22%), transpiration rate (37.84%), stomatal conductance (25.54%), water use efficiency (50.87%), intrinsic water use efficiency (11.58%) and intercellular CO_2 (5.86%) as compared to well water control.

Many studies have shown the decreased photosynthetic activity under drought stress due to stomatal or non-stomatal mechanisms (Del Blanco *et al.*, 2000; Samarah *et al.*, 2009). Stomata are the entrance of water loss and CO_2 absorbability and stomatal closure is one of the first responses to drought stress which result in declined rate of photosynthesis. Stomatal closure deprives the leaves of CO_2 and photosynthetic carbon assimilation is decreased in favor of photorespiration. Considering the past literature as well as the current information on drought-induced photosynthetic responses, it is evident that stomata close progressively with increased drought stress. It is well known that leaf water status always interacts with stomatal conductance and a good correlation between leaf water potential and stomatal conductance always exists, even under drought stress. It is now clear that there is a drought-induced root-to-leaf signaling, which is promoted by soil drying through the transpiration stream, resulting in stomatal closure. The "non-stomatal" mechanisms include changes in chlorophyll synthesis, functional and structural changes in chloroplasts, and disturbances in processes of accumulation, transport, and distribution of assimilates.

With respect to the effect of drought on *chlorophyll content*; chlorophyll is one of the major chloroplast components for photosynthesis, and relative chlorophyll content has a positive relationship with photosynthetic rate. The decrease in chlorophyll content under drought stress has been considered a typical symptom of oxidative stress and may be the result of pigment photo-oxidation and chlorophyll degradation. Photosynthetic pigments are important to plants mainly for harvesting light and production of reducing powers. Both the chlorophyll a and b are prone to soil dehydration (Farooq *et al.*, 2009). Decreased or unchanged chlorophyll level during drought stress has been reported in many species, depending on the duration and severity of drought (Kpyoarissis *et al.*, 1995; Zhang and Kirkham, 1996). Drought stress caused a large decline in the chlorophyll a content, the chlorophyll b content, and the total chlorophyll content in different sunflower varieties (Manivannan *et al.*, 2007b).

Loss of chlorophyll contents under water stress is considered a main cause of inactivation of photosynthesis. Furthermore, water deficit induced reduction in chlorophyll content has been ascribed to loss of chloroplast membranes, excessive swelling, distortion of the lamellae vesiculation, and the appearance of lipid droplets (Kaiser *et al.,* 1981). Low concentrations of photosynthetic pigments can directly limit photosynthetic potential and hence primary production. From a physiological perspective, leaf chlorophyll content is a parameter of significant interest in its own right. Studies by majority of chlorophyll loss in plants in response to water deficit occurs in the mesophyll cells with a lesser amount being lost from the bundle sheath cells.

In terms of the role of drought in the accumulation of osmolytes: plants accumulate different types of organic and inorganic solutes in the cytosol to lower osmotic potential thereby maintaining cell turgor (Rhodes and Samaras, 1994). Under drought, the maintenance of leaf turgor may also be achieved by the way of osmotic adjustment in response to the accumulation of proline, sucrose, soluble carbohydrates, glycinebetaine, and other solutes in cytoplasm improving water uptake from drying soil. The process of accumulation of such solutes under drought stress is known as osmotic adjustment which strongly depends on the rate of plant water stress. Wheat is marked by low level of these compatible solutes and the accumulation and mobilization of proline was observed to enhance tolerance to water stress (Nayyar and Walia, 2003). Of these solutes, proline is the most widely studied because of its considerable importance in the stress tolerance. Proline accumulation is the first response of plants exposed to water-deficit stress in order to reduce injury to cells. Progressive drought stress induced a considerable accumulation of proline in water stressed maize plants. The proline content increase as the drought stress progressed and reached a peak as recorded after 10 day stress, and then decreased under severe water stress as observed after 15 days of stress (Anjum *et al.,* 2011b).

Proline can act as a signaling molecule to modulate mitochondrial functions, influence cell proliferation or cell death and trigger specific gene expression, which can be essential for plant recovery from stress (Szabados and Savoure´, 2009). Accumulation of proline under stress in many plant species has been correlated with stress tolerance, and its concentration has been shown to be generally higher in stress-tolerant than in stress-sensitive plants. It influences protein solvation and preserves the quaternary structure of complex proteins, maintains membrane integrity under dehydration stress and reduces oxidation of lipid membranes or photoinhibition

(Demiral and Turkan, 2004). Furthermore, it also contributes to stabilizing sub-cellular structures, scavenging free radicals, and buffering cellular redox potential under stress conditions (Ashraf and Foolad, 2007).

CHAPTER TWO

2.1   Definition of Metabolite

Plant metabolism is defined as the complex of physical and chemical events of photosynthesis, respiration and the synthesis and degradation of organic compounds (Anonymous, 2018). Photosynthesis produces the substrates for respiration and the starting organic compounds used as building blocks for subsequent biosynthesis of nucleic acids, amino-acids, and proteins, carbohydrates and organic acids, lipids and natural products. Plants undergo both primary and secondary metabolism. In fact, while secondary metabolism facilitates the primary metabolism in plants; the primary metabolism consists of chemical reactions that allow the plant to live.

Metabolites are the intermediates and products of metabolism. The term *metabolite* is usually restricted to small molecules. Metabolites have various functions, including fuel, structure, signaling, stimulatory and inhibitory effects on enzymes, catalytic activity of their own (usually as a cofactor to an enzyme), defense, and interactions with other organisms (Tiwari and Rana, 2015).

It is important to note here that in order for the plants to stay healthy, secondary metabolism plays a pivotal role in keeping all the plants' systems working properly. A common role of secondary metabolites in plants is defense mechanisms.

2.1.1   *Examples of Metabolite*

According to Buchanan *et al.* (2015), primary metabolites includes lactic acid, certain amino acids while examples of some of secondary metabolites present include peptides / growth factors, antibiotics, terpenoids, alkaloids, nucleosides among others.

2.2   Types of Metabolite

Plants require both primary and secondary metabolism for their optimum functioning.

2.2.1 *Primary metabolite*

Plant primary metabolism in a plant comprises all metabolic pathways that are essential to the plant's survival. Meanwhile, primary metabolites are compounds that are directly involved in the growth and development of a plant.

2.2.2 *Secondary metabolite*

Secondary plant metabolites are produced by other metabolic pathways, although important are not essential to the functioning of the plant. They are used in signaling and regulation of primary metabolic pathways. For instance; plant hormones which are secondary metabolites are often used to regulate the metabolic activity within the cells and oversee the overall development of the plant.

CHAPTER THREE

3.1   Plant Physiology and Metabolism

Plants have the most sophisticated chemical system in the world (Stitt *et al.*, 2010). They use light energy to convert CO_2 into carbohydrates in their leaves. They absorb nutrients like nitrate, phosphate, and sulfate via their roots and convert them to amino acids and nucleotides, using light energy in the leaves in the day and energy derived from respiration in leaves in the dark and in non-photosynthetic tissues. Carbohydrates, amino acids, and nucleotides are then transported to growing tissues, where they are converted into macromolecular cellular components like proteins, nucleic acids, cell walls, pigments, and lipids.

Plants also synthesize a number of secondary metabolites, including phenylpropanoids and flavonoids, terpenoids, glucosinolates, and alkaloids (Stitt *et al.*, 2010). These have important roles in cellular function, in signaling, and in adaptation to abiotic and biotic stress. Their unique synthetic ability is the result of a highly complex and sophisticated metabolic apparatus. Previous report has pointed out the complexity and flexibility of plants as revealed in the discoveries of several pathways like glycolysis, the oxidative pentose phosphate pathway, and organic acid metabolism, present in more than one compartment (Lunn, 2007). This is in addition to the unveiling of the diversity of plant secondary metabolite.

Plants metabolites are known to be unique sources for pharmaceuticals food additives, flavors and others industrial values (Tiwari and Rana, 2015).

3.2 Role of Secondary metabolites

Plants possess several natural products. These Natural products are those chemical compounds or substances that are isolated from living organism. It can be in form of primary or secondary metabolites.

It has been well-established that the primary metabolite is directly involved in normal Growth, development and reproduction. For example, carbohydrate, protein, fat and oil, alcohol etc.
On the other hand, secondary metabolites are not directly involved in growth, development and reproduction of an organism, but they have an ecological function (Nwokeji *et al.*, 2016). Plant secondary metabolite can be found in the leaves, stem, root or the bark of the plant depending on the type of secondary metabolite that is been produced (Hill, 1952; Nwokeji *et al.*, 2016). Unlike the primary metabolites, absence of secondary metabolites does not result in immediate death, but rather in long term impairment of the organisms' survivability (Nwokeji *et al.*, 2016).

Although from various sources, plant secondary metabolites are generally classified into three distinct groups namely; Terpenes, Phenolic compounds and Nitrogen-containing compounds. Their biosynthesis are known derived from primary metabolism pathways, which include tricarboxylic acid cycle (TCA), methylerithrotol phosphate (MEP) pathway, mevalonic and shikimic acid pathway.

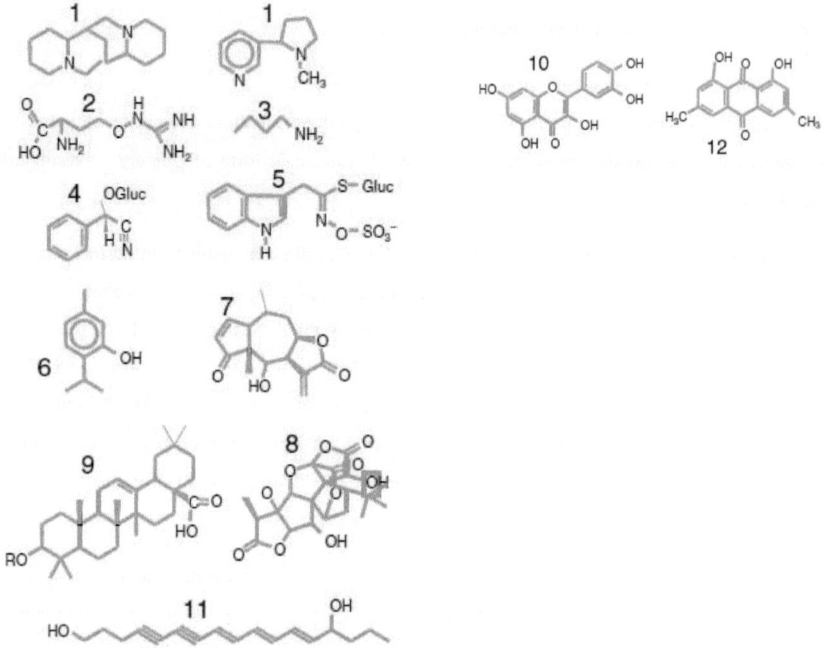

Figure 3.1: With N [1. Alkaloids; 2. Non-protein amino acids; 3.Amines; 4.Cyanogenic glycosides; 5.Glucosinolates]; Without N [6.Monoterpenes; 7.Sesquiterpenes; 8. Diterpenes; 9.Triterpenes, Saponins, Steroids, Tetraterpenes; 10.Flavonoids; 11.Polyacetylenes; 12.Polyketides, Phenylpropanes]

3.2.1 *Terpenes*

This constitutes the largest class of secondary product. They are also called terpenoids. The diverse substances of this class are generally insoluble in water. All terpenes are derived from the union of five-carbon atoms that have the branched carbon skeleton of isopentane. The basic structural element of terpenes is sometimes called isoprene unit because terpene decompose at high temperature to give isoprene. Terpene are toxic and feeding deterrents to many plants feeding insect and mammals. Thus they appear to have important defensive role in the plant kingdom (Gershenzon and Croteau, 2012).

Derivatives of terpene called saponins are steroid and triterpene glycoside, so named because of their soap like properties. Another derivative of terpene is carotenoids which give the yellow, red and orange colour in some plants like carrot. Examples of plant that contain terpenoids include *polypodiumvulgare*, *Digitalis spp,* pine and fir, peppermint plant, lemon, basil, sage,corn,*Gossypiumhispida*(cotton),wild tobacco (Kessler and Baldwin, 2001).

3.2.2 *Phenolic Compounds*

Plants produce a large variety of secondary product that contains a phenol group- a hydroxyl functional group on an aromatic ring. These substances are classified as phenolic compounds. Plant phenolic are a chemically heterogeous compound, some soluble only in organic solvents, some are water soluble, while others are insoluble polymers. Some simple phenolic are activated by ultra violet light. Phenolic are wide spread in vascular plants and appear to function in different capacities. The derivatives of phenolic compounds include simple phenyl propanoid, benzoic acid derivatives, anthocyannin, isoflavones, tannins, lignin, and flavonoid compound beginning with phenylalanine. Lignin is generally formed from three different phenyl propanoid alcohols namely coniferyl, coumaryl, and sinapyl (Davin and Lewis, 2005). The flavonoids are one of the largest classes of plant phenolics, the basic structure contain 15 carbon arranged in two aromatic ring connected by a three carbon bridge (Taiz and Zeiger, 2005). The basic function of the flavonoids is for pigmentation and defence. The red, pink, purple and blue colours observed in plants parts are as a result of anthocyanins (Li *et al.*, 2003). Tannins were first used to describe compound that could convert raw material hides into leather in the process of tanning. Tannins are generally toxic that significantly reduce the growth and survivorship of many herbivores when added to their diets. Tannins can be seen in fruits like apple, black berries, tea and red wine (Taiz and Zeiger, 2005). Tannins are mainly constituent of woody plants especially heart wood. Some derivatives of tannin include Gallic acid.

3.2.3 *Nitrogen-containing compounds*

A large variety of secondary metabolites have nitrogen in their structure. These include the alkaloids, cyanogenic glucoside, glucosinate (Taiz and Zeiger, 2005).

Alkaloids are large family of more than 15,000 nitrogen containing secondary metabolites found in approximately 20% of the species of vascular plant. The nitrogen atom in these substances is usually part of the heterocyclic ring, a ring that contain both nitrogen and carbon atom. They show striking pharmacological effect on vertebrate animal.as their name would suggest. Most alkaloids are alkaline, at pH value commonly (7.2). Alkaloids were once thought to be nitrogenous wastes. Most alkaloids are now believed to function as defense against especially mammals, because of the general toxicity and deterrence capacity (Hartmann, 2013). One group of alkaloid, the pyrrolizidine alkaloid illustrates how herbivore can become adapted to tolerate plant defensive substance and even use them in their own defence (Hartmann, 2013).

In addition to the classes of secondary metabolites; it is important to consider the biosynthesis of secondary metabolites.

Consequently, the biosynthesis of the various three classes of secondary metabolite varies depending on the class involved. The terpenes are biosynthesized from primary metabolite through the pathway of mevalonic acid and methyerithrotol phosphate (Taiz and Zeiger, 2005). In the mevalonic acid pathway, three molecules of acetyl-coA are joined together stepwise to form mevalonic acid. This key six-carbon intermediate is then pyrophosphorylated, decarboxylated and dehydrated to yield isopentenyldiphosphate (IPP). The IPP is the activated five carbon building block of terpenes. Although all the details have not yet been elucidated, glyceraldehyde-3-phosphate and two carbon atoms derived from pyruvate appear to combine to generate an intermediate that is eventually converted to IPP.

In biosynthesis of phenolic compound, two basic pathways are involved, the shikimic acid pathway and the malonic acid pathway (Taiz and Zeiger, 2005). The shikimic acid pathway convert simple carbohydrate precursors derived from glycolysis and the pentose phosphate pathway to the aromatic amino acid (Hartmann, 2013). One of the intermediate is shikimic acid, which has given its name to this whole sequence of reaction. The well-known broad spectrum herbicide *glyphosphate* kill plant by blocking a step in this pathway.

The most abundant classes of secondary phenolic compound in plant are derived from phenylalanine via the elimination of an ammonia molecule to form cinnamic acid (Taiz and Zeiger, 2005). This reaction is catalyzed by phenylalanine ammonia lyase (PHL) perhaps the most studied enzyme in plant secondary metabolite.

For the nitrogenous compound, the pathway is the formation of aliphatic amino acid through the TCA cycle.

Secondary metabolites are also known for their economic importance including

- Plant breeding (Levison, 2006)
- Biological potential [e.g. Anti-Microbial] (Okwu, 2001; Oliver, 2006; Duke, 2012)
- Health and Pharmaceutical, in the formulation of various drugs like antibiotics, analgesics (Lewis and Elvin-Lewis, 1995).
- Plant Defence against Herbivores and Pathogens (Wink, 1998).
- Toxicity (e.g. the toxicity potential of secondary metabolites like strychnine, cyanogenic glycosides, coumarinsetc).
- Allelopathic Effect [for instance; simple phenyl propanoids and benzoic acid derivatives are frequently cited as having allelopathicactivites] (Taiz and Zeiger, 2005).

CHAPTER FOUR

4.1   Antioxidant system and Plant metabolism

Under normal conditions, free radicals and ROS generation occurs at a low level and its generation and removal are balanced. This balance is disturbed by increasing ROS level due to the environmental stress factors (Sharma *et al.*, 2012). For the removal of excess ROS and reducing oxidative damages, plants have evolved anti-oxidative defense systems consisting of non-enzymatic and enzymatic mechanisms.

The **Non-enzymatic antioxidants** include major cellular redox buffers ascorbate (vitamin A) and glutathione (Apel and Hirt, 2004). Tocopherol (vitamin E), flavonoids, caretonoids and phenolics are also components of non-enzymatic antioxidant system. They take place in defense systems as well as influence plant growth and development. **Enzymatic ROS scavenging system** consists of several antioxidants enzymes including superoxide dismutase (SOD), catalase (CAT), enzymes of ascorbate-glutathione cycle being ascorbate peroxide (APX), monodehydroascorbatereductase (MDHAR), dehydroascorbatereductase (DHAR) and glutathione reductase (GR) (Sharma *et al.*, 2012).

- *Non-enzymatic Antioxidants*

Among the non-enzymatic antioxidants, ascorbate is the most abundant one. It buffers cell against oxidative damage of high ROS level. It is synthesized in mitochondria and transferred to other cellular compartments including chloroplast as well as in apoplast. Due to the ability to donate electrons, ascorbate is a powerful antioxidant (Sharma *et al.*, 2012). Ascorbate protects membrane by directly reacting with O_2^- and H_2O_2 and also takes role in removal of H_2O_2 via ascorbate-glutathione cycle (Foyer *et al.*, 1997) that is shown in Figure 1.7-c.

Under environmental stress factors, level of ascorbate depends on the balance between ascorbate synthesis rate and turnover related to antioxidant demand (Chaves *et al.*, 2002). Stress tolerant plants induce overexpression of enzymes related ascorbate synthesis.

The tripeptide glutathione is one of the major redox buffers in aerobic cells (Foyer *et al.*, 2001). It is a low molecular weight nonproteinthiol and an important part of the antioxidative defense system. GSH is synthesized in cytosol and chloroplast and it is transferred to different cellular compartments (Sharma *et al.*, 2012). Due to its reducing power, GSH has many roles in different biological processes such as cell growth, signal transduction, enzymatic regulation, protein synthesis and expression of the stress-related genes (Foyer *et al.*, 1997).

As an antioxidant, GSH takes part in ascorbate-glutathione cycle, as well as reacting with hydrogen peroxide to be oxidized to GSSG (Desikan*et al.*, 2003). Maintenance of reduced GSH is vital for the cell. Under environmental stress conditions the GSH/GSSG ratio is altered, promoting the GSH synthesizing enzyme's activity (Vanacker*et al.*, 2000).

Tocopherols are another type of antioxidant involved in ROS scavenging. They present only green parts of plants. Tocopherols protect membrane components including lipids via reacting with O_2 in chloroplast (Ivanov and Khorobrykh, 2003). They also avoid chain propagation of lipid auto-oxidation.

Carotenoids belonging to the group of lipophilic antioxidants, are able to detoxify ROS (Young, 1991). Carotenoids inhibit oxidative damage via removing $^1O_2$ and also prevent $^1O_2$ formation by quenching triplet and excited chlorophyll to protect photosynthetic system. Besides ROS scavenging roles, carotenoids take place in signaling to enhance stress responses.

Phenolic compounds, such as flavanoids, esters, lignin and tannins are secondary metabolites found in plant tissues. They have variety of functions as antioxidants including, removal of reactive oxygen species, preventing lipid peroxidation, chelating transition metal ions and decreasing membrane fluidity. These processes limit peroxidation via hindering the ROS diffusion into the cells (Arora *et al.*, 2000).

- ***Enzymatic Anti-oxidative Defense Systems***

The enzymatic ROS scavenge systems consists of superoxide dismutase (SOD), catalase (CAT) and enzymes of ascorbate-glutathione cycle (APX, MDHAR, DHAR). Although these enzymes function in different cell compartments, they work in collaboration as responding to ROS damage.

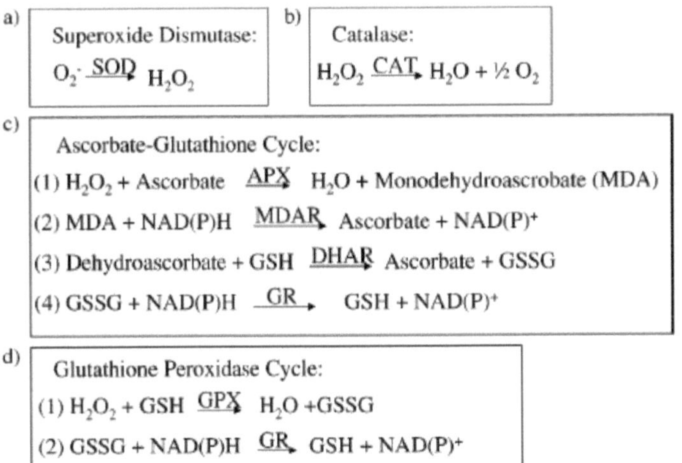

Figure 4.1: Enzymatic ROS scavenge mechanisms (Apel and Hirt, 2004)

SOD, being a metalloenzyme found mainly in three isoforms in plants. The isozymes are classified according to the metal co-factors of the enzyme and they operate at different parts of the cell. Manganese SOD functions in mitochondria, while iron SOD present in chloroplast and cupper/zinc SOD found in cytosol, chloroplast, peroxisome and mitochondria (Jackson *et al.*, 1978). SOD catalyses the dismutation of superoxide to oxygen and hydrogen peroxide. As a result of environmental stresses, SOD activity of cells increases as a tolerance mechanism. High levels of SOD activity is an indicator of resistance to the stress factor (Zaefyzadeh*et al.*, 2009).

Catalase is a ubiquitous, tetrameric, heme-containing enzyme (Sharma *et al.*, 2012). It has high specificity for hydrogen peroxide and catalyzes the degradation of hydrogen peroxide to water and oxygen as shown in -b. Catalase is located mainly in peroxisomes, where is the major cellular compartment of H_2O_2 synthesis via photorespiratory oxidation and β-oxidaiton of fatty acids (Corpas*et al.*, 2008).

Ascorbate-glutathione cycle is an important regulator of the oxidative balance of cells. The AsA-GSH cycle consists of detoxification of H_2O_2 via the interactions of ascorbate peroxidase (APX), monodehydroascorbatereductase (MDHAR), dehydroascorbatereductase (DHAR) and glutathione reductase (GR) (Desikan*et al.*, 2003).

Ascorbate peroxidase is a heme peroxidase and has an important role for balancing ROS level, as an AsA-GSH cycle member. It catalyzes the reduction of H_2O_2 to water by using two molecules of ascorbate. As an end product of this reaction, monodehydroascorbate (MDHA) is generated (Welinder, 1992; Patterson and Poulos, 1995). The MDHA radical is converted to ascorbate via MDHAR enzyme using NADPH as electron donor (Sakiham*et al.*, 2000). Although ascorbate is regenerated from MDHA by enzymatic reactions, an amount of DHA is produced during the oxidation ofascorbate. This DHA is also reduced to ascorbate via DHAR enzyme oxidizing GSH to GSSG (Ushimaru*et al.*, 1997). In order to maintain the cellular GSH/GSSG ratio, glutathione reductase, a flavoenzyme, regenerate GSH from GSSG.

To remove reactive oxygen species and eliminate their oxidative damage, the balance between the anti-oxidative enzymes is very important. Overexpression of one component could not be sufficient for the defense, while enhancing combination of enzymes has been shown to increase tolerance (Aono*et al.*, 1995; Kwon *et al.*, 2002).

4.2   Environmental stressors

The environment affects an organism in many ways, at any time. They affect the general well-being of the organism. These so-called environmental factors can be categorized based on their abiotic and biotic nature.

Biotic environmental factors; includes interactions with other organisms resulting in infection or mechanical damage by herbivory or trampling, as well as effects of symbiosis or parasitism. On the other hand, ***abiotic environmental factors*** include temperature, humidity, light intensity, the supply of water and minerals, and CO_2; these are the parameters and resources that determine the growth of a plant (Blum, 1986; Mahajan and Tuteya, 2005; Dita*et al.*, 2006).

BIOTIC STRESSES	ABIOTIC STRESSES
1. Viruses 2. Bacteria 3. Insects 4. Herbivores 5. Rodent 6. Weeds	1. Extreme temperatures (low & high) 2. Drought 3. Flooding 4. Salinity 5. Heavy metals 6. Pollutants 7. Oxidative stress 8. Soil structure (nutrient deprivation) 9. Extreme wind 10. Radiation

Table 4.1: Various Biotic and Abiotic Stress Factors

It is hard to measure the exact force applied by stress and also a condition could be a stress factor for one plant while it is an optimum condition for another plant. Thus, it is difficult to define stress in biological terms (Mahajan and Tuteya, 2005). However, for practical purposes, biological stress can be defined as an overpowering pressure of some adverse force or condition that inhibits normal functions, growth and well-being of biological systems (Jones *et al.*, 1989).

Plant growth and productivity is adversely affected by various abiotic and biotic stress factors. Plants are frequently exposed to a plethora of stress conditions such as low temperature, salinity, drought, flooding, heat, oxidative stress and heavy metal toxicity. Various anthropogenic activities have further accentuated the existing stress factors (Aksoy, 2008; Mahajan and Tuteya, 2005).

In response to these stress factors, plants have developed many stress tolerance mechanisms. These mechanisms may vary among species at different developmental stages (Ashraf and Wu, 1994), although basic responses to stress factors are conserved among most of the plant species (Zhu, 2001). In addition, different stress factors may lead to similar responsive adaptations like up-regulating the stress proteins and increasing compatible solute accumulation (Zhu, 2002).

Briefly, it is important to state that all stress tolerance mechanisms are initiated by sensing the stress signals via the interaction of the extracellular materials with a plasma membrane protein. Following the perception of the signal, secondary signals are generated immediately (Agarwal and Zhu, 2005). Changes in the level of these secondary signals include calcium, inositolphospates (IPs) and reactive oxygen species (ROS), up-regulates further signals. Each secondary signal initiates a phosphorylation cascade, which triggers the expression of stress responsive genes and the transcription factors of these genes (Mahajan and Tuteya, 2005; Agarwal and Zhu, 2005). The stress responsive genes produce various osmolytes, antioxidants, proteins functioning in stress tolerance.

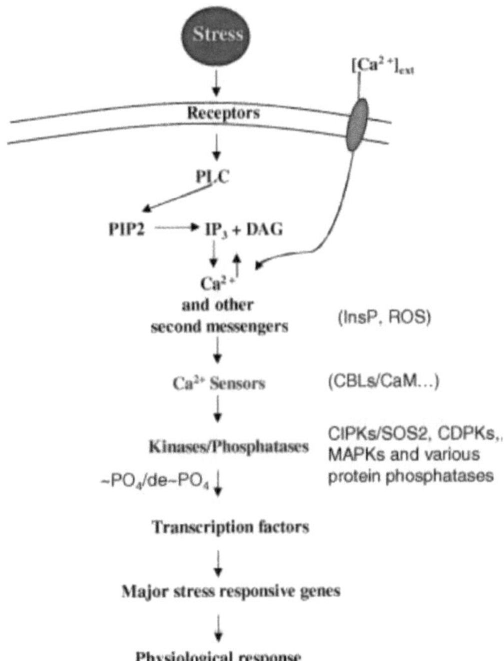

Figure 4.2: Signal Transduction Pathway in response to Abiotic Stress (Mahajan and Tuteya, 2005)

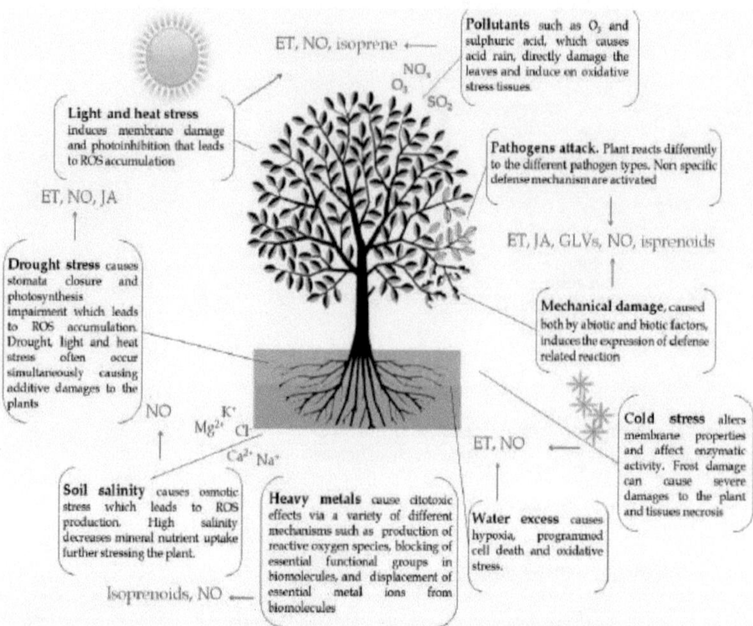

Figure 4.3: Consequences of stresses and some plant physiologic responses. ET = ethylene, ROS = Reactive oxygen species, NO = nitric oxide; JA = Jasmonic acid; GLVs = Green Leaf Volatile (s); O$_3$ = Ozone (Wang and Frei, 2011; Kadioglu*et al.*, 2012).

4.3 Free radicals / Reactive Oxygen Species (ROS)

Various mechanisms are known concerning the impact of environmental stresses to plant metabolism. In spite of this, one underlining factor to the role of stress on plant physiology is that of free radicals and reactive oxygen species (ROS).

Oxygen is necessary for the normal growth of plants, because of aerobic processes such as photosynthesis and cellular respiration. It is also implicated in leads to the production of reactive oxygen species (ROS) in mitochondria, chloroplast and peroxisomes. All ROS types have the potential to cause oxidative damage to lipids, proteins and DNA (Apel and Hirt, 2004).

In plants, ROS are produced continuously as byproducts of different metabolic pathways (Elster, 1991). The main source of ROS in plants is the photosynthetic electron transport system. There are two major processes involved in the generation of ROS during photosynthesis. One is the direct photoreduction of oxygen to superoxide radical by photosystem I (PSI) electron transport chain. The other one is the oxygenase reaction of rubisco taken place in photorespiratory pathway (Arora*et al.,* 2002; Apel and Hirt, 2004). By these reactions, molecular oxygen is converted to superoxide by the removal of single electron. From this superoxide anion, hydrogen peroxide (H_2O_2) and hydroxyl radicals are formed via series of reductions (Apel and Hirt, 2004; Agarwal and Zhu, 2004).

Under non-stressful conditions, the production and removal of the ROS are controlled by various anti-oxidative defense mechanisms and plants are protected against harmful effects of these active oxygen molecules. However, the equilibrium between production and elimination of ROS is disturbed by many abiotic stress factors resulting in rapid rising of the cellular level of ROS.

ROS are also thought to be functioning as signaling molecules in defense response pathways of plants. Among the reactive oxygen species, H_2O_2 is more likely to be a signaling molecule, since its half-life is longer than the other ROS, it is uncharged and able to diffuse through aqueous and lipid phases (Apel and Hirt, 2004; Agarwal and Zhu, 2004).

Singlet Oxygen

Singlet oxygen is the electronically excited state of the molecular oxygen and less stable than the molecular oxygen. It destructs biological molecules by reacting with them.

The chlorophyll pigments, which are the components of photosynthetic reaction center, are the main source of the singlet oxygen ($^1O_2$). It is generated during the triplet chlorophyll production, in PSII.

Superoxide

A superoxide is formed when oxygen is reduced by a single electron, during the mitochondrial electron transport chain or during photosynthesis. During photosynthesis, ferredoxin or the electron carriers on the reducing side of PSI donates their electrons to oxygen forming superoxide radical, O_2^-. It is thought that most of the superoxide anions are produced by the reduced ferredoxin (Arora*et al.*, 2002).

Throughout mitochondrial electron transport chain, molecular oxygen is reduced to superoxide anion either in the flavaprotein region of NADH dehydrogenase or in the ubiquinone-cytochrome region, as seen in the Figure 1.5 (Arora*et al.*, 2002).

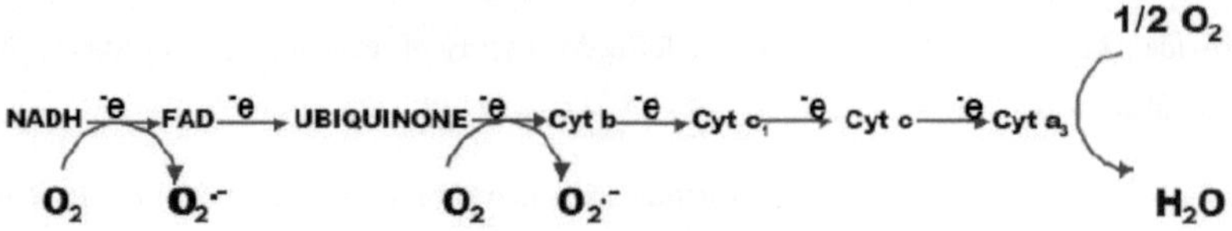

Figure 4.4: Superoxide Formation sites in mitochondrial electron transport chain. (Arora*et al.*, 2002).

Since its extra electron is unpaired, superoxide is a free radical and relatively unstable, so that it is either converted back to the molecular oxygen or further reduced to H_2O_2.

Hydrogen Peroxide

Hydrogen peroxide (H_2O_2) is a product of normal metabolism taking place in peroxisome, chloroplast and electron transport chain in mitochondria. It acts both as an oxidant and as a reductant. Hydrogen peroxide is produced by the dismutation of superoxide and hdyroperoxy radical (HO_2^-) (Upadhyaya*et al.*, 2007; Aksoy, 2008).

Various environmental stresses induce hydrogen peroxide production via enzymes including NADPH oxidases localized on plasma membrane and cell wall peroxidases (Neill *et al.*, 2002). Besides the normal metabolism, H_2O_2 can be generated by superoxide dismutases (SOD). Different types of SOD present at different locations in the cell, such as iron-containing SOD (FeSOd) being in chloroplast and manganase-containing SOD (MnSOD) being in mitochondria. Besides being a toxic oxygen species, hydrogen peroxide functions as a signaling factor in stress signaling pathways. It initiates localized oxidative damage in leaf cells and changes the redox status of the surrounding to start anti-oxidative response.

Hydroxyl Radicals

Among the reactive oxygen species, hydroxyl radicals are the most damaging ones. Although hydrogen peroxide and superoxide radical do not directly destruct the vital cellular components like DNA, proteins and plasma membranes; they generate the damaging hydroxyl radicals. Hydroxyl radicals are produced according to the Haber-Weiss reaction in the presence of ferric ion, which is summarized as;

$$H_2O_2 + O_2^{\cdot -} \xrightarrow{\ Fe^{2}, Fe^{3+}\ } OH^{\cdot} + OH^- + O_2$$

Hydroxyl radicals destruct organic substances via oxidation, either by the addition of OH to the molecule or by the abstraction of an H atom from the molecule (Arora*et al.*, 2002).

With respect to ROS and Oxidative Damage to biomolecules present in plants;
Under normal conditions, production and removal of ROS are strictly controlled. When the level of ROS exceeds the defense mechanisms, organism is said to be under oxidative stress. Increased level of ROS causes various damages to biological molecules that can be seen below (Sharma *et al.*, 2012).

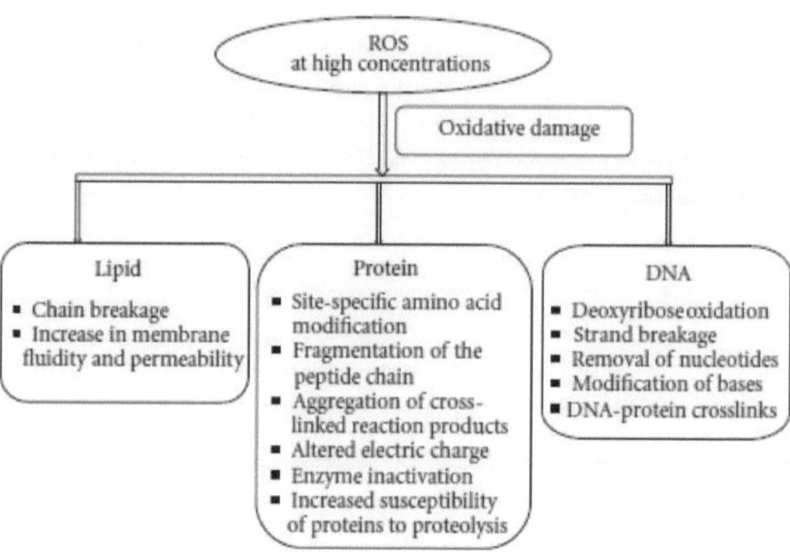

Figure 4.6: Oxidative damage of ROS on lipids, proteins and DNA (Sharma *et al.*, 2012).

When ROS levels increases, lipid peroxidation is triggered in cellular and organellar membranes. Lipid-derived radicals, which are produced as a result of lipid peroxidation, increases oxidative stress via reacting with proteins and DNA (Han *et al.*, 2009; Mishra *et al.*, 2011; Sharma *et al.*, 2012).

Malondialdehyde (MDA) being one of the end-products of phospholipid peroxidation is responsible for the membrane damage (Halliwell and Gutteridge 1989).

On phospholipid molecules there are two main sites for the ROS attack; the double (unsaturated) bond between two carbon atoms and the ester linkage. Thus, polyunsaturated fatty acids are more vulnerable to the ROS attacks. Lipid peroxidation process consists of three stages as initiation, progression and termination (Smirnoff, 1995). Peroxidation of phospholipids ends up with many reactive species such as aldehydes, lipid epoxides, alcohols, alkoxyl radicals and alkanes and leads to the increase in membrane permeability (Sharma *et al.*, 2012).

Alterations of proteins upon the ROS attacks can be either direct or indirect. Protein activity modulation via carbonylation, nitrosylation or disulphide bond formation constitute directalteration, while indirect modification occurs by the interaction with end-products of lipid peroxidation (Yamauchi *et al.*, 2008). High levels of ROS lead to the site-specific amino acid modification, peptide-chain fragmentation, increase in proteolysis susceptibility and charge alterations (Moller and Kristensen, 2004).

Amino acids of a protein have different vulnerability to ROS attacks. Thiol groups and iron-sulphur centers of sulphur-containing amino acids are the most vulnerable sites for ROS attack. oxidized peptides increases the proteolytic digestions (Cabiscol *et al.*, 2000).

ROS are also responsible for the DNA damages. They act by causing oxidative damage all types of DNA; nuclear, mitochondrial and chloroplastic. Since mitochondrial and chloroplast DNA lack repair systems, they are more susceptible to oxidative damages than the nuclear DNA (Richter, 1992). Although nuclear DNA has repair system, excess ROS leads to permanent damages to DNA that mostly result in changes at protein level ending up with malfunctioning or complete inactivation of proteins. Some of the damages of ROS attack on DNA are strand breakage, deoxyribose oxidation, nucleotide removal and modifications or removal of nucleotides (organic base part) that further results in mismatches with the other strand (Sharma *et al.*, 2012).

Oxidative attacks on bases of DNA occur via OH addition to the double bonds, while sugar damages occur as a result of hydrogen removal from the deoxyribose (Dizdaroğlu, 1993). $^1O_2$ reacts only with guanine base, on the other hand H_2O_2 and O_2^- do not react any of the bases (Dizdaroğlu, 1993; Halliwell and Aruoma, 1991).

Oxidative damages on the DNA sugars results in single-strand breakage. Attack of ROS produces deoxyribose radical via removal of hydrogen atom from the C4' position of the sugar, which in turn generate strand breakage (Evans *et al.*, 2004).

The hydroxyl radical attacks on the DNA and related proteins lead to the DNA-protein cross-links, which can be lethal if replication or transcription takes place before repair system activation.

4.4   Effect of Drought on Plant Physiology and Metabolism

All living organisms have two fundamental natures, which are the cellular organization and requirement for liquid water (Wood, 2007). In plants, water has many functions. Water accounts for 80% - 95% of fresh weight of non-woody plants, being the main medium for transporting metabolites and nutrients. It is also the major solvent with its unique biophysical properties including high heat of vaporization and high surface tension. Due to these properties water can remain liquid over a wide temperature range and solvate many molecules.

Water has roles in a number of biochemical processes as a reactant like being electron donor. Besides these biochemical functions, water is the key component in maintaining cell turgor (Wood, 2007; Bartels and Souer, 2004).

Nevertheless, studies have shown that water can be a source of stress. Water stress has been linked with excess of water or water deficit (Mahajan and Tuteja, 2005).

Excess of water results in reduced oxygen in roots, which in turns results in disruption of root functions such as respiration and nutrient uptake. The more common water stress is water deficit, which is called as drought. Drought is the limitation of water over a prolonged period of time.

In crop production worldwide, water deficiency is the main limiting factor (Mahajan and Tuteja, 2005).

The challenge of drought continues to be a source of concern since time immemorial. It is a regular and severe constriction to crop yields in many areas of the world where lentils were grown (McWilliam, 1986).

- ***Impact of Drought Stress on Cell Integrity and Plant Growth***

As a result of water removal from the cell membrane, lipid bilayer structure of the membrane is disrupted and membrane proteins are displaced. This leads to lose of membrane integrity, selectivity and interruption of cellular compartmentalization. Due to the intense water deficit, cells shrink and mechanical strain occurs on membranes. All these defects damage the functioning of transporters, ions and membrane based enzymes (Mahajan and Tuteja, 2005). As a consequence of cell shrinkage cellular volume decreases, resulting in viscous cellular content that increases protein aggregation and denaturation via protein-protein interaction (Hoekstra *et al.*, 2001).

Another effect of water deficit is the reduction of vegetative growth. Under drought stress conditions cyclin-dependent kinase activity reduces, resulting in slower cell division and even inhibition of growth (Shuppler*et al.*, 1998). Leaf growth is more sensitive than the root growth to water deficiency, as reducing leaf area is advantageous for plants decreasing water loss through transpiration under drought conditions (Mahajan and Tuteja, 2005).

- ***Effect of Drought Stress on Photosynthesis***

The rate of photosynthesis decreases due to the stomatal closure, under water deficit conditions. Photosynthetic system in plants depends on the availability of CO_2, especially in photosystem II. Under drought stress, the primary reason of the decline in photosynthetic rate is the CO_2 deficiency (Meyer *et al.*, 1998). The closure of stomata under drought stress leads to the decrease in intracellular CO_2 levels, which in turns results in over-reduction of electron transport chain components. Thus, the electrons are transferred to oxygen at photosystem I generating reactive oxygen species (Mahajan and Tutja, 2005).

Water deficiency also results in decreasing Rubisco, a carboxylatingenzyme, activity thus limits photosynthesis (Bota*et al.*, 2004). In plants, the amount of rubisco is controlled by the rate of synthesis and degradation (Reddy *et al.*, 2004). Under drought stress conditions, synthesis of rubisco decreases.

Normally, rubiscoactivase regulates the active site conformation of rubisco, removes inhibitors allowing the enzyme to undergo carboxylation (Chaves *et al.*, 2002). During water deficiency, rubiscoactivase activity decreases due to the reduced ATP concentrations. Thus, removal of inhibitors from rubisco active site is impaired (Tezara *et al.*, 1999).

4.5 *Relationship between ROS and Drought Stress*

Under drought stress conditions, production of reactive oxygen species is known to increase in several ways. For instance; down regulation of photosystem II due to the water deficiency, results in an imbalance between generation and consumption of electrons. The changes occurring in photosystem II results in the dissipation of excess light energy generating reactive oxygen species including O_2^-, $^1O^+$, H_2O_2 and OH (Peltzer *et al.*, 2002).

Superoxide radicals are also generated due to the changes in the photosynthetic electron transport chain under drought stress.

Also, inhibition of CO_2 assimilation, coupled with the changes in photosystem I and II and electron transport chain result in enhanced ROS production (Asada, 1999). During water deficiency stomatal closure results in reduced CO_2 fixation that leads to reduction in $NADP^+$ production via Calvin cycle. Thus the lack of electron acceptor results in overproduction of electrons through photosynthetic electron transport chain and these electrons are trapped by O_2, generating ROS (Sharma *et al.*, 2012).

Oxidative damages occurring during drought stress are due to the overproduction of reactive oxygen species. ROS attack the most important cellular components to disrupt their function. Some of the ROS dependent damages are amino acid and protein oxidation, DNA nicking and lipid peroxidation (Asada, 1999). If the damaged components of the cell are not repaired, the cell death occurs, eventually.

4.6 Effect of Salinity on Plant Physiology and Metabolism

The soil is the most important factors for the plants growth, mainly governed by weathering and climatic conditions. Soil is formed from weathering of rocks. The rate of plants growth and type of species depend upon the mineral composition of soil and parent rock. Plant nutrients are mainly derived from weathering of minerals. The mineralogical composition studies have importance in forestry where the plants growth lasts over a long period and depends to a large extent on the mineral as a source of plant nutrients in the soil.

According to a survey, more than 800 million hectares of land throughout the world are affected by stress caused by imbalances in soil salt concentration (Anonymous, 2008).

Salt stress is a serious problem globally (Moghbeli *et al.*, 2012); such stress retards plant growth and yield. Additionally, this stress is one of the most important abiotic stresses in arid and semiarid regions (Olfati *et al.*, 2012; Talebi *et al.*, 2015).

Rengasamy (2010) pointed that saline soils are mainly dominated by Na^+ and Cl^- anions, in which NaCl constitutes from 50 to 80% of the total soluble salts.

Salinity (also described as soil nutrient) is an abiotic stress, and can negatively affect the plant's morphological characteristics as well as the quality and quantity of its phytochemical compounds of, including total phenol, total soluble sugars and its components, namely sucrose, glucose, and fructose (Asghari and Ahmadvand, 2018).

The impact of salt stress has been correlated with some morphological and physiological traits such as reduction in fresh and dry weight (Chartzoulakis and Klapaki, 2000). In fact, salinity affects plant metabolism by disturbing their physiological and biochemical processes of plants by inducing ionic and osmotic imbalances which slows down plant growth and productivity (Munns, 2005).

The deleterious effects of salinity on plant growth are associated with low osmotic potential of soil solution, nutritional imbalance, specific ion effect, or a combination of these factors (Ashraf and Harris, 2004).

Mechanisms of action caused by salt stress on plant cells arise from the following:
(1) Disruption of ionic equilibrium (Influx of Na+ dissipates the membrane potential and facilitates the uptake of Cl^- down the chemical gradient).
(2) Na^+ is toxic to cell metabolism and has deleterious effect on the functioning of some of the enzymes.
(3) High concentrations of Na^+ causes osmotic imbalance, membrane disorganization, reduction in growth, inhibition of cell division and expansion. induction of ion imbalance in cells, especially lowering concentrations of K^+, Ca^{2+}, and NO^{3-}; and causing ion (Na^+ and/or Cl^-) toxicity.
(4) High Na^+ levels also lead to reduction in photosynthesis and production of reactive oxygen species (Mahajan and Tuteja, 2005; Tavakkoli, 2011).

4.7   Relationship between Water- and Salt- stresses

Salinity in a given land area depends upon various factors like amount of evaporation (leading to increase in salt concentration), or the amount of precipitation (leading to decrease in salt concentration).

Inland deserts are marked by high salinity as the rate of evaporation far exceeds the rate of precipitation. Agricultural lands that have been heavily irrigated are highly saline. As drier areas in particular need intense irrigation, there is extensive water loss through a combination of both evaporation as well as transpiration. This process is known as evapotranspiration and as a result, the salt delivered along with the irrigation water gets concentrated, year-by-year in the soil. This leads to huge losses in terms of arable land and productivity as most of the economically important crop species are very sensitive to soil salinity.

Examples of salt sensitive plants, also known as glycophytes include rice (*Oryza sativa*), maize (*Zea mays*), soybean (*Glycine max*) and beans (*Phaseolus vulgaris*).

High salt concentration (Na^+) in particular which deposit in the soil can alter the basic texture of the soil resulting in decreased soil porosity and consequently reduced soil aeration and water conductance.

As illustrated above, the basic physiology of high salt stress and drought stress overlaps with each other. High salt depositions in the soil generate a low water potential zone in the soil making it increasingly difficult for the plant to acquire both water as well as nutrients. Therefore, salt stress essentially results in a water deficit conditions in the plant and takes the form of a physiological drought.

The major ions involved in salt stress signaling, include Na^+, K^+, H^+ and Ca^{2+}. It is the interactions of these ions, which brings homeostasis in the cell.

4.8 Effect of drought and salinity on Secondary Metabolites

- *Drought*

  Apart from soil water availability is also one of the important factors in plants growth and development. Magnitude of drought and period are usually documented in many environments and can severely impact on plant's stress tolerance and survival. Photosynthetic rate is reduced under less availability of soil water or drought condition. As of result of reduction of photosynthetic rate, the rate of plants growth is also reduced.

To this effect; Alonso-Amelot *et al.* (2007) reported that phenolic biosynthesis in high temperature and reduced water availability. Biosynthesis of secondary metabolites is dependent on primary metabolism. On the other hand, primary metabolism is effected by photosynthetic rate.

During reduced water availability, plants close their stomata and restrict photosynthesis and therefore one might expect negative relationship between biosynthesis of secondary metabolites and water shortage. That why phenolic and saponin contents and their pharmacological activity are reported to vary seasonally in medicinal bulbs (Ncube *et al.*, 2011a).

It is well established that the accumulation of natural products strongly depends on the growing conditions, such as the temperature, the light regime and the nutrient supply (Falk *et al.*, 2007, Ballhorn *et al.* 2011). In addition, more severe environmental influences, such as various stress conditions, will also impact on the metabolic pathways responsible for the accumulation of secondary plant products (Bohnert *et al.*, 1995). Unfortunately, to date, only limited information on this complex issue has been made available.

A wide range of experiments have shown that plants exposed to drought stress did indeed accumulate higher concentrations of secondary metabolites. Such enhancement is reported to occur in nearly all classes of natural products, such as simple or complex phenols, numerous terpenes, as well as in nitrogen-containing substances, such as alkaloids, cyanogenic glucosides or glucosinolates (Selmar and Kleinwachter, 2012).

Selmar and Kleinwachter (2012) however argues that although drought stress frequently enhances the concentration of secondary plant products, the stress-related increase in natural product concentrations indeed does not mean that the rate of biosynthesis of natural products in the plants has increased, since the drought stress applied also reduces growth and biomass production in most plants. Consequently, a simple and, at first sight, obvious and rational explanation for this effect is that: 'Plants suffering drought stress, in principle, synthesize and accumulate the same amounts of natural products as well-watered conditions but, due to the reduction in overall plant biomass, product concentration on a fresh or dry weight basis is enhanced!'

- *Salt*

Induced biosynthesis of phenolic compounds is reported in soil contains low level of iron (Dixon and Paiva, 1995). Meanwhile, increased pro-anthocyanidins contents were reported in soil contains limited phosphate.

Soil nutrients are directly linked with biosynthesis of secondary metabolites as well as their abundance or limited availability has effect on contents of secondary metabolites in plants (Kouki and Manetas, 2002).

To this effect, there are several hypotheses with soil nutrients. Bryant *et al.* (1983) proposed the carbon/nitrogen balance (CNB) hypothesis and postulated that fertilization in presence of limited available nutrients will cause low production of carbon-based secondary metabolites.

According to the CNB hypothesis contents of secondary metabolites deposition in plant tissues depends on relative excess of plant resources, especially nitrogen. Generally, the primary metabolism is necessary for plants growth and development and in priority for plants over secondary metabolism, after fulfillment of requirement of carbon and nitrogen in primary metabolism plants growth, they are allocated for secondary metabolites production. However, intermediates of primary metabolism are precursor of secondary metabolism and precursor contents determine the production rate of secondary metabolites (Reichardt et al., 1991). In deficiency of nitrogen in plants, CNB predicts that carbohydrates will accumulate in plant tissues and stimulate the production carbon-based secondary compounds.

To this effect, it has been argued that the accumulation of carbon-based/N-based compounds depends on the availability of both elements. In deficiency of nitrogen elements, biosynthesis of C-based compounds increases and vice versa. Nitrogen containing compounds like cyanogenic glycosides and alkaloids contents was reported low in soil of nitrogen deficiencies and results of more experiments has found consistent with these predictions (Simon *et al.*, 2010), yet CNB also unsuccessful repeatedly in other similar studies (Riipi *et al.*, 2002).

Jones and Hartley (1999) have also given protein competition model (PCM) of phenolic distribution in plants, to support the CNB theory. According to protein competition model (PCM) production of phenolic compounds depend on constituent derived from metabolic pathways or pathways precursors and are regulated by biochemical mechanism. Diallinas & Kanellis (1994) have reported that biosynthesis of alkaloid and phenolic compounds are catalysed by Phenylalanine ammonia-lyase (PAL) and utilize the amino acid phenylalanine as a precursor.

In addition, previous findings revealed that there is competition between protein and metabolite synthesis for the limiting phenylalanine and leading to process level exchange among rate of protein versus metabolite synthesis. Jones & Hartley (1999) hence concluded in terms of synthesis rate, stating that if rate of protein synthesis is high, invariably the rate of phenolic synthesis should be low and vice versa.

Finding of these two theories (CNB and PCM) clearly revealed that rate of secondary metabolite biosynthesis is consequence of the interaction between the plant's intrinsic (genetic) and extrinsic (environmental) factors.

Conclusion(s)

Plants in nature are continuously exposed to several biotic and abiotic stresses. Among these stresses, drought and salinity occupies clearly a non-negligible position, causing the most adverse effects on plant growth and productivity.

Various studies have also established that extrinsic factor (biotic and abiotic stresses) induces biosynthesis of secondary metabolites in spite of the conflicting trends for different classes of secondary metabolites. It is thus pertinent to understand holistically how these environmental stresses impact the availability of secondary metabolites and how these metabolites in turn affect the health status of the plant. Furthermore, researchers have emphasized that immeasurable potentials exist in employing such stresses as a tool to improve the quality as well as pharmacological properties of plant material as a whole.

References

Aksoy, E. (2008). "Effect of drought and salt stresses on the gene expression levels of antioxidant enzymes in lentil (*Lens culinaris M.)* seedlings". MSc Thesis. Middle East Technical University, Ankara Turkey.

Alam MZ, Stuchbury T, Naylor REL, Rashid MA (2004) Effect of salinity on growth of some mordern rice cultivars. J. Agron. 3:1-10.

Alonso-Amelot, M.E., Oliveros-Bastidas, A., Calcagno-Pisarelli, M.P. (2007).Phenolics and condensed tannins of high altitude *Pteridiumarachnoideum*in relation to sunlight exposure, elevation, and rain regime.*Biochem. Syst. Ecol.* 35, 1–10.

American Meteorological Society (2003). Meteorological Drought, Policy statement. (Available online at:) http:// www.ametsoc.org.

Anjum S.A., Xie X., Wang L., Saleem M.F., Man C (2011). Morphological, physiological and biochemical responses of plants to drought stress. African Journal of Agricultural Research Vol. 6(9), pp. 2026-2032. DOI: 10.5897/AJAR10.027.

Anjum SA, Wang LC, Farooq M, Khan I, Xue LL (2011b). Methyl jasmonate-induced alteration in lipid peroxidation, antioxidative defense system and yield in soybean under drought. J. Agron. Crop Sci., doi:10.1111/j.1439-037X.2010.00468.x.

Aono, M., H. Saji, et al. (1995)."Paraquat Tolerance of Transgenic Nicotianatabacum with Enhanced Activities of Glutathione Reductase and Superoxide Dismutase." Plant and Cell Physiology 36(8): 1687-1691.

Apel, K. and H. Hirt (2004). "Reactive Oxygen Species: Metabolism, Oxidative Stress, and Signal Transduction." Annual Review of Plant Biology 55(1): 373-399.

Arora, A., T. M. Byrem, et al. (2000). "Modulation of liposomal membrane fluidity by flavonoids and isoflavonoids." Arch BiochemBiophys373(1): 102-109.

Asada, K. (1999). "The water-water cycle in chloroplasts: scavenging of active oxygen and dissipation of e cess photons." Ann Rev Plant Physiol Plant MolBiol50: 601–39.

Ashraf M, Foolad MR (2007). Roles of glycinebetaine and proline in improving plant abiotic stress resistance. Environ. Exp. Bot., 59: 206-216.

Ashraf M, Waheed A (1990) Screening of local/exotic accessions of lentil (Lens culinaris Medic) for salt tolerance at two growth stages. Plant and Soil 128:167–176.

Ashraf, M. and L. Wu (1994)."Breeding for Salinity Tolerance in Plants." Critical Reviews in Plant Sciences 13(1): 17-42.

Ashraf, M., & Harris, P. J. C. (2004). Potential biochemical indicators of salinity tolerance in Plants. *Plant Science, 166*(1), 3-16.

Bado S., Forster B.P., Ghanim A.M.A., Jankowicz-Cieslak J., Berthold L., Luxiang L. (2016). Protocols for Pre-Field Screening of Mutants for Salt Tolerance in Rice, Wheat and Barley. International Atomic Energy Agency. Pp 1 - 44. ISBN: 978-3-319-26590-2. DOI 10.1007/978-3-319-26590-2.

Ballhorn, D.J., Kautz, S., Jensen, M., Schmitt, S., Heil, M. and Hegeman, A.D. (2011). Genetic and environmental interactions determine plant defences against herbivores. J. Ecol. 99: 313–326.

Ballhorn, D.J., Kautz, S., Jensen, M., Schmitt, S., Heil, M. and Hegeman, A.D. (2011) Genetic and environmental interactions determine plant defences against herbivores. J. Ecol. 99: 313–326.

Bartels, D. and E. Souer (2004).Molecular responses of higher plants to dehydration.Plant Responses to Abiotic Stress. H. Hirt and K. Shinozaki, Springer Berlin / Heidelberg. 4: 9-38.

Benjamin JG, Nielsen DC (2006). Water deficit effects on root distribution of soybean, field pea and chickpea. Field Crops Res., 97: 248-253.

Blum, A., R. Munns, et al. (1996). "Genetically engineered plants resistant to soil drying and salt stress. How to interpret osmotic relations?" Plant Physio110: 1051.

Bohnert, H.J., Nelson, D.E. and Jensen, R.G. (1995).Adaptations to environmental stresses. Plant Cell 7: 1099–1111.
Bota, J., H. Medrano, et al. (2004). "Is photosynthesis limited by decreased Rubisco activity and RuBP content under progressive water stress?" New Phytologist. 162(3): 671-681.

Bray E.A., Bailey-Serres J., Weretilnyk E. (2000). Responses to abiotic stresses, in: W. Gruissem, B. Buchannan, R. Jones (Eds.), Biochemistry and Molecular Biology of Plants, American Society of Plant Biologists, Rockville, MD, pp. 158–1249.

Buchanan B.B., Gruissem W., Jones R.L. (2015). Biochemistry and Molecular Biology of Plants. John Wiley & Sons. ISBN 9781118502198.

Cabiscol, E., E. Piulats, et al. (2000). "Oxidative Stress Promotes Specific Protein Damage in Saccharomyces cerevisiae." Journal of Biological Chemistry 275(35): 27393-27398.

Chandra S (2003). Effects of leaf age on transpiration and energy exchange of *Ficus glomerata,* a multipurpose tree species of central Himalayas. Physiol. Mol. Biol. Plants, 9: 255-260.

Chartzoulakis K., Klapaki, G. (2000). Response of two greenhouse pepper hybrids to NaCl salinity during different growth stages. *Science Horticulture, 86*(3), 247-260.

Chartzoulakis, K., &Klapaki, G. (2000).Response of two greenhouse pepper hybrids to NaCl salinity during different growth stages.*Science Horticulture, 86*(3), 247-260.

Chaves, M. M., J. S. Pereira, et al. (2002). "How Plants Cope with Water Stress in the Field? Photosynthesis and Growth." Annals of Botany. 89(7): 907-916.
Coley, P.D., Bryant, J.P., Chapin, S.F. (1985).Resource availability and plant anti-herbivore defence.*Science,* 230, 895–899.

Corpas, F. J., J. M. Palma, et al. (2008). "Peroxisomal xanthine oxidoreductase: characterization of the enzyme from pea (Pisumsativum L.) leaves." J Plant Physiol165(13): 1319-1330.

Davin, L. B. and Lewis, N. G. Dirigent protein and dirigent sites explain the mystery of specificity of radical precursor coupling in lignan and lignin biosynthesis. Plant Physiology 123: pp 453-461. 2000.

De Abreu CEB, Prisco JT, Nogueira ARC, Bezerra MA, de Lacerda CF, Filho EG (2008) Physiological and biochemical changes occurring in dwarf-cashew seedlings subjected to salt stress. Braz. J. Plant Physiol. 20:105-118.

Demiral T, Turkan I (2004). Does exogenous glycinebetaine affect antioxidative system of rice seedlings under NaCl treatment? J. Plant Physiol., 161: 1089-1110.

Demirevska K, Zasheva D, Dimitrov R, Simova-Stoilova L, Stamenova M, Feller U (2009). Drought stress effects on Rubisco in wheat: changes in the Rubisco large subunit. Acta Physiol. Plant., 31: 1129-1138.

Desikan, R., J. Hancock, et al. (2004). Oxidative stress signaling Plant Responses to Abiotic Stress. H. Hirt and K. Shinozaki, Springer Berlin / Heidelberg. 4: 121-149.

Dewy DR (1962) Breeding crested wheatgrass for salt tolerance. Crop Sci 2:403–407.

Dixon, R.A., Paiva, N.L. (1995). Stress-induced phenylpropanoid metabolism.*The Plant Cell*, 7, 1085–1097.

Dizdaroglu, M. (1993). Chemistry of free radical damage to DNA and nucleoproteins. DNA and Free Radicals, B. Halliwell and O. I. Aruoma, Eds. Ellis Horwood, London, UK: 19–39.

Duan B, Yang Y, Lu Y, Korpelainen H, Berninger F, Li C (2007). Interactions between drought stress, ABA and genotypes in Picea asperata. J. Exp. Bot., 58: 3025-3036.

Duke, A. Pyrrole pigment isoprenoid compound and phenolic plant constituents.Vol. 9. Elsevier, New York, U .S. A. 960p. 2012.

Epstein E, Norlyn JD (1977) Seawater-based crop production: a feasibility study. Science 197:247–261.

Epstein E, Norlyn JD, Rush DW, Kingsbury RW, Kelly DB, Cunningham GA, Wrona AF (1980) Saline culture of crops: a genetic approach. Science 210:399–404.

Evans, M. D., M. Dizdaroglu, et al. (2004). "Oxidative DNA damage and disease: induction, repair and significance." Mutation research 567(1): 1-61.

Falk, K.L., Tokuhisa, J.G. and Gershenzon, J. (2007) The effect of sulfur nutrition on plant glucosinolate content: physiology and molecular mechanisms. Plant Biol. 9: 573–581.

Farooq M, Wahid A, Kobayashi N, Fujita D, Basra SMA (2009). Plant drought stress: effects, mechanisms and management. Agron. Sustain. Dev., 29: 185-212.

Fessenden, R. J. and Fessenden, J. S. Organic Chemistry (2nd edn.). Willard Grant Press, Massachusetts, U. S. A. 57p. 1997.

Flowers TJ (2004). Improving crop salt tolerance. J Exp Bot 55:307–319.

Foyer, C. H., H. Lopez-Delgado, et al. (1997). "Hydrogen peroxide- and glutathione-associated mechanisms of acclimatory stress tolerance and signalling." Physiologia Plantarum100(2): 241-254.

Garthwaite AJ, von Bothmer R, Colmer TD (2005) Salt tolerance in wild *Hordeum*species is associated with restricted entry of Na+ and Cl– into the shoots. J. Exp. Bot. 56:2365-2378.

Gershenzon, J. and Croteau, R. Terpenoid in Herbivores: Their interaction with Secondary Plant Metabolite: The Chemical participant. Vol.1(2nd edn.). Academic Press San Diego, California, U. S. A. 409 p. 2012.

Gleadow, R.M. and Woodrow, I.E. (2002). Defense chemistry of cyanogenic Eucalyptus cladocalyx seedlings is affected by water supply. Tree Physiol. 22: 939–945.

Gregorio GB, Senadhira D, Mendoza RT (1997) Screening rice for salinity tolerance, vol 22, IRRI discussion paper series. IRRI, Manila, p 30.

Halliwell, B. and O. I. Aruoma (1991)."DNA damage by oxygen-derived species.Its mechanism and measurement in mammalian systems." FEBS letters 281(1-2): 9-19.

Han, C., Q. Liu, et al. (2009) "Short-term effects of experimental warming and enhanced ultraviolet-B radiation on photosynthesis and antioxidant defense of *Piceaasperata*Seedlings." Plant Growth Regulation 58(2):153–162.

Harris D, Tripathi RS, Joshi A (2002). On-farm seed priming to improve crop establishment and yield in dry direct-seeded rice, in: Pandey S., Mortimer M., Wade L., Tuong T.P., Lopes K., Hardy B. (Eds.), Direct seeding: Research Strategies and Opportunities, International Research Institute, Manila, Philippines, pp. 231-240.

Hartmann, T. Chemical ecology of pyrrolizidine alkaloids. Planta207: pp483-495. 2013.

Hill, A. F. Economic Botany. A textbook of useful plant and plant Products (2nd edn.). MC-Graw-Hill Book Company Inc., New York, U .S .A. 743p. 1952.

Hoekstra, F. A., E. A. Golovina, et al. (2001). "Mechanisms of plant desiccation tolerance." Trends in plant science 6(9): 431-438.

Horvath, E., Szalai, G., Janda, T. (2007). Induction of abiotic stress tolerance by salicylic acid signaling. *Journal of Plant Growth Regulation, 26*(3), 290-300.

Hosseini SM., Sharifzadeh A., Akbari M (2009). Causes, Effects and Management Mechanisms of Drought Crisis in Rural and Nomadic Communities in Southeastern Iran as Perceived by Agricultural/Rural Managers and Specialist. J Hum Ecol, 27(3): 189-200.

Hussain M, Malik MA, Farooq M, Ashraf MY, Cheema MA (2008). Improving drought tolerance by exogenous application of glycinebetaine and salicylic acid in sunflower. J. Agron. Crop Sci., 194: 193-199.

Inderjit, P., Dakshini, K. M. and Emhellia, F. A. Allelopathy Organism, Processes and Application ACS Symposium Series American Chemical Society, Washington, DC.27p. 1995.

Ivanov, B. N. and S. Khorobrykh (2003)."Participation of photosynthetic electron transport in production and scavenging of reactive o ygen species." Antioxidants and Redox Signaling 5(1): 43–53.

J G Ambede, G W Netondo, G N Mwai, D M Musyimi **(2012).**NaCl salinity affects germination, growth, physiology, and biochemistry of bambara groundnut. **Brazilian Society of Plant Physiology.**24(3): 151-160.

Jackson, C., J. Dench, et al. (1978). "Subcellular Localisation and Identification of Superoxide Dismutase in the Leaves of Higher Plants." European Journal of Biochemistry 91(2): 339-344.

Johns, T. Chemical Ecology of the Aymara of Western Bolivia: Selection for Glyco-alkaloids in the Solanum X Ajanhuiri Domestication Complex.PhD Dissertation, University of Michigan, Ann Arbor, U .S .A. 65p. 1985.

Jones H.G., Jones M.B. (1989). Introduction: some terminology and common mechanisms, in: H.G. Jones, T.J. Flowers, M.B. Jones (Eds.), Plants Under Stress, Cambridge university Press, Cambridge, pp. 1–10.

Jones, C.G., Hartley, S.E. (1999). A protein competition model of phenolic allocation.*Oikos*, 86, 27–44.

Kaiser WM, Kaiser G, Schöner S, Neimanis S (1981). Photosynthesis under osmotic stress. Differential recovery of photosynthetic activities of stroma enzymes, intact chloroplasts and leaf slices after exposure to high solute concentrations. Planta, 153: 430-435.

Kamara AY, Menkir A, Badu-apraku B, Ibikunle O (2003). The influence of drought stress on growth, yield and yield components of selected maize genotypes. J. Agric. Sci., 141: 43-50.

Kessler, A. and Baldwin, I. T. Defensive function of herbivore Induced plant volatile emission in nature. Science 29: pp2141-2144. 2001.

Khan MB, Hussain N, Iqbal M (2001). Effect of water stress on growth and yield components of maize variety YHS 202. J. Res. (Science), 12: 15-18.

Kouki, M., Manetas, Y. (2002). Resource availability affects differentially the levels of gallotannins and condensed tannins in Ceratoniasiliqua. *Biochemical Systematics and Ecology* ,30, 631–639.

Kwon, S. Y., Y. J. Jeong, et al. (2002). "Enhanced tolerances of transgenic tobacco plants expressing both superoxide dismutase and ascorbate peroxidase in chloroplasts against methyl viologen-mediated oxidative stress." Plant, Cell & Environment 25(7): 873-882.

Kyparissis A, Petropoulun Y, Manetas Y (1995). Summer survival of leaves in a soft-leaved shrub (*Phlomis fruticosa* L., Labiatae) under Mediterranean field conditions: avoidance of photoinhibitory damage through decreased chlorophyll contents. J. Exp. Bot., 46: 1825-1831.

Levison, H. Z (2006). The defensive role of alkaloids in insects and plants. Experienta 92: pp 408-411.

Lewis, W. H. and Elvin-Lewis, M. P. Medicinal plants as source of new therapeutics. Annals of Missouri Botanic Garden 82: pp16-24. 1995.

Li, J., Lee, O., Raba, R. and Last, R . L. Arabidopsis flavonoid mutant are hypersensitive to UV-B radiation. Plant Cell 85: pp171-179. 2003.

Mahajan, S. and N. Tuteja (2005). "Cold, salinity and drought stresses: an overview." Arch BiochemBiophys444(2): 139-158.

Manivannan P, Jaleel CA, Kishorekumar A, Sankar B, Somasundaram R, Sridharan R, Panneerselvam R (2007a). Changes in antioxidant metabolism of *Vigna unguiculata* L. Walp. by propiconazole under water deficit stress. Colloids Surf B: Biointerf., 57: 69-74.

Manivannan P, Jaleel CA, Sankar B, Kishorekumar A, Somasundaram R, Alagu Lakshmanan GM, Panneerselvam R (2007b). Growth, biochemical modifications and proline metabolism in *Helianthus annuus* L. as induced by drought stress. Colloids Surf. B: Biointerf., 59: 141-149.

McWilliam, J. R. (1986). "The national and international importance of drought and salinity effects on agricultural production." Australian Journal of Plant Physiology 13: 1–13.

Meyer, S. and B. Genty (1998). "Mapping intercellular CO_2 mole fraction (Ci) in rosarubiginosa leaves fed with abscisic acid by using chlorophyll fluorescence imaging. Significance Of ci estimated from leaf gas exchange." Plant Physiol116(3): 947-957.

Mishra, S., A. B. Jha, et al. (2011). "Arsenite treatment induces o idative stress, upregulates antioxidant system, and causes phytochelatin synthesis in rice seedlings." Protoplasma248(3): 565–577.

Mogadas-Farimani S, Hossaini SM 2004. Pasture ecosystem management in drought. *Proceeding of Agricultural Science in Arid Area.* Ardestan Azad University press, Iran (in Persian).

Moghbeli, E., Fathollahi, S., Salari, H., Ahmadi, G., Saliqehdar, F., Safari, A., &Hosseini, G. M. (2012).Effects of salinity stress on growth and yield of *Aloe Vera* L. *Journal of Medicinal Plants Research, 6*(16), 3272-3277.

Moller, I. M. and B. K. Kristensen (2004)."Protein oxidation in plant mitochondria as a stress indicator." Photochemical &Photobiological Sciences 3(8): 730-735.

Munns R, Tester M (2008) Mechanisms of salinity tolerance. Annu. Rev. Plant Biol. 59:651-681.

Munns, R. (2005). Genes and salt tolerance: bringing them together. *New Phytologist, 167*(3), 645-663.

Musyimi DM (2005) Evaluation of young avocado plants (*Perseaamericana*Mill.) for tolerance to soil salinity by physiological parameters. Maseno University, Kenya. MSc thesis.

Nayyar H, Walia DP (2003). Water stress induced proline accumulation in contrasting wheat genotypes as affected by calcium and abscisic acid. Biol. Plant., 46: 275-279.

Ncube, B., Finnie, J.F., Van Staden, J. (2011a). Seasonal variation in antimicrobial and phytochemical properties of frequently used medicinal bulbous plants from South Africa. *South African Journal of Botany*, 77, 387–396.

Neill, S., R. Desikan, et al. (2002)."Hydrogen pero ide signalling," Current Opinion in Plant Biology 5(5): 388–395.

Nonami H (1998). Plant water relations and control of cell elongation at low water potentials. J. Plant Res., 111: 373-382.

Nwokeji P A, Enodiana O I, Ezenweani S R, Osaro-Itota O, Akatah H A (2016). The Chemistry of Natural Product: Plant Secondary Metabolites. International Journal of Technology Enhancements and Emerging Engineering Research, Vol 4, Issue 8.

Okwu, D. E. Evaluation of the chemical composition of Indigenous species and flavoring agents. Global Journal of Pure and Applied Science 7(3): pp455-459. 2001.

Olfati, J. A., Moqbeli, E., Fathollah, S., &Estaji, A. (2012). Salinity stress effects changed during *Aloe Vera* L. vegetative growth. *Journal of Stress Physiology and Biochemistry, 8*(2), 152-158.

Oliver, B. Medicinal plant in tropical West Africa. CamBridge University press, New York.155p. 2006.

Owens T, Hoddinott J, Kinsey B (2003). Ex-ante Actions and Ex-post Public Responses to Drought Shocks: Evidence and Simulations from Zimbabwe. *World Development,* 31(7): 1239–1255.

Pakniyat H, Handley LL, Thomas WTB, Connolly T, Macaulay M, Caligari PDS, Forster BP (1997) Comparison of shoot dry weight, Na+ content and δ13C values of ari-e and other semi dwarf barley mutants under salt stress. Euphytica 94:7–14.

Patterson, W. R. and T. L. Poulos (1995)."Crystal structure of recombinant pea cytosolic ascorbate peroxidase." Biochemistry 34(13): 4331-4341.

Peltzer, D., E. Dreyer, et al. (2002). "Temperature dependencies of antioxidative enzymes in two contrasting species." Plant PhysiolBiochem40:141–50.

Plant Responses to Abiotic Stress. H. Hirt and K. Shinozaki, Springer Berlin / Heidelberg. 4: 121-149.

Praba ML, Cairns JE, Babu RC, Lafitte HR (2009). Identification of physiological traits underlying cultivar differences in drought tolerance in rice and wheat. J. Agron. Crop Sci., 195: 30-46.

R Asghari, R Ahmadvand (2018). Salinity Stress and its impact on Morpho-Physiological Characteristics of *Aloe Vera*.Pertanika J. Trop. Agric. Sci. 41 (1): 411 – 422.

Reichardt, P.B., Chapin III, F.S., Bryant, J.P., Mattes, B.R., Clausen, T.P. (1991). Carbon/nitrogen balance as a predictor to of plant defence in Alaskan balsam poplar: potential importance of metabolite turnover. *Oecologia*, 88,401–406.

Rengasamy P (2010) Soil processes affecting crop production in salt-affected soils. Aust. J. Soil Res. 37:613-620.

Rhodes D. Samaras Y (1994). Genetic control of osmoregulation in plants. In cellular and molecular physiology of cell volume regulation. Strange, K. Boca Raton: CRC Press, pp. 347-361.

Richards RA, Dennett CW, Qualset CO, Epstein E, Norlyn JD, Winslow MD (1987) Variation in yield of grain and biomass in wheat, barley and triticale in a salt-affected field. Field Crops Res 15:277–278.

Riipi, M., Ossipov, V., Lempa, K., Haukioja, E., Koricheva, J., Ossipova, S., Pihlaja, K. (2002).Seasonal changes in birch leaf chemistry: are there trade-offs between leaf growth and accumulation of phenolics? *Oecologia*, 130, 380–390.

Rucker KS, Kvien CK, Holbrook CC, Hook JE (1995). Identification of peanut genotypes with improved drought avoidance traits. Peanut Sci., 24: 14-18.

Rush DW, Epstein E (1976). Genotypic response to salinity: differences between salt sensitive and salt tolerant genotypes in the tomato. Plant Physiol 57:162–166.

Schuppler, U., P. H. He, et al. (1998)."Effect of water stress on cell division and cell-division-cycle 2-like cell-cycle kinase activity in wheat leaves." Plant Physiol117(2): 667-678.

Shannon MC (1985) Principles and strategies in breeding for higher salt tolerance. Plant Soil 89:227–241.

Sharma, P. and A. B. Jha, et al. (2012)."Reactive Oxygen Species, Oxidative Damage, and Antioxidative Defense Mechanism in Plants under Stressful Conditions." Journal of Botany 2012: 26.

Simon, J., Gleadow, R.M., Woodrow, I.E. (2010). Allocation of nitrogen to chemical defence and plant functional traits is constrained by soil. *N. Tree Physiology,* 30, 1111–1117.

Smirnoff, N. (1998). "Plant resistance to environmental stress."CurrOpinBiotechnol9(2): 214-219.

Szabados L, Savoure′ A (2009). Proline: a multifunctional amino acid. Trends Plant Sci., 15: 89-97.

Taiz, L. and Zeiger, E. Plant physiology.Third Edition.Sinauer Association Inc., California, U .S. A. 690p. 2005.

Talebi, S., Jafarpour, M., Mohammadkhani, A., &Sadeghi, A. (2012).The effect of different concentrations of salicylic acid and sodium chloride on Iranian Borage.*International Journal of Agricultural Crop Science, 4*(18), 1348-1352.

Tavakkoli E, Fatehi F, Coventry S, Rengasamy P, McDonald GK (2011) Additive effects of Na+ and Cl- ions on barley growth under salinity stress. J. Exp. Bot. 62:2189-2203.

Tezara, W., V. J. Mitchell, et al. (1999). "Stress inhibits plant photosynthesis by decreasing coupling factor and ATP." Nature 1401:914–7.

Tiwari R., Rana C.S. (2015). Plant secondary metabolites: a review. International Journal of Engineering Research and General Science Volume 3, Issue 5, September-October, 2015 ISSN 2091-2730.

Upadhyaya, H., M. H. Khan, et al.(2007)." Hydrogen peroxide induces oxidative stress in detached leaves of *Oryza sativa* L." Gen. Appl. Plant Physiology 33(1-2): 83-95.

Ushimaru, T., Y. Maki, et al. (1997). "Induction of Enzymes Involved in the Ascorbate-Dependent Antioxidative System, Namely, Ascorbate Peroxidase, MonodehydroascorbateReductase and DehydroascorbateReductase, after Exposure to Air of Rice (Oryza sativa) Seedlings Germinated under Water." Plant and Cell Physiology 38(5): 541-549.

Vanacker, H., T. L. W. Carver, et al. (2000). "Early H2O2 accumulation in mesophyll cells leads to induction of glutathione during the hypersensitive response in the barley-powdery mildew interaction." Plant Physiol. 123:1289–300.

Welinder, K. G. (1992). "Superfamily of plant, fungal and bacterial peroidases." Current Opinion in Structural Biology 2(3): 388–393.

Wink, M. Plant Breeding: Importance of plant secondary metabolite for protection against pathogen and herbivores Theoretical and Applied Genetics 75: pp225-233. 1998.

Wood, A. J. (2007). Eco-physiological Adaptations to Limited Water Environments. Plant Abiotic Stress, Blackwell Publishing Ltd: 1-13.

Yamauchi, Y., A. Furutera, et al. (2008). "Malondialdehyde generated from peroxidizedlinolenic acid causes protein modification in heat-stressed plants." Plant PhysiolBiochem. 46(8-9): 786-793.

Young, Y. (1991). "The photoprotective role of carotenoids in higher Plants."PhysiologiaPlantarum. 83(4): 702–708.

Zamani GH, Gorgievski-Duijvesteijn MJ, Zarafshani K (2006). Coping with Drought: Towards a Multilevel Understanding Based on Conservation of Resources Theory. *Hum Ecol,* 34: 677–692.

Zhao TJ, Sun S, Liu Y, Liu JM, Liu Q, Yan YB, Zhou HM (2006). Regulating the drought-responsive element (DRE)-mediated signaling pathway by synergic functions of trans-active and transinactive DRE binding factors in *Brassica napus*. J. Biol. Chem., 281: 10752-10759.

Zhu, J. K. (2001). "Plant salt tolerance." Trends Plant Sci6(2): 66-71.

Zhu, J. K. (2002). "Salt and drought stress signal transduction in plants." Annual Review Plant Biology 53: 247-273.